Living Together

Please visit our website at: www.garethstevens.com
For a free color catalog describing Gareth Stevens'
list of high-quality books and multimedia programs,
call 1-800-542-2595 (USA) or 1-800-461-9120 (Canada).
Gareth Stevens Publishing's Fax: (414) 332-3567.

Library of Congress Cataloging-in-Publication Data available upon request from publisher. Fax: (414) 336-0157 for the attention of the Publishing Records Department.

ISBN 0-8368-2817-8

This North American edition first published in 2001 by
Gareth Stevens Publishing
330 West Olive Street, Suite 100
Milwaukee, WI 53212 USA

© QA International, 2000

Created and produced as *So Many Ways to Live in Society* by
 QA INTERNATIONAL
329 rue de la Commune Ouest, 3ᵉ étage
Montréal, Québec
Canada H2Y 2E1
Tel.: (514) 499-3000 Fax: (514) 499-3010
www.qa-international.com

Printed in Canada

1 2 3 4 5 6 7 8 9 05 04 03 02 01

Gareth Stevens Publishing
A WORLD ALMANAC EDUCATION GROUP COMPANY

Strength in numbers!

No human being can live completely alone. We need each other to feed and house ourselves and to care for our children. Animals also gather together. Some do so temporarily. Others stay together throughout their lives. Some form simple groups. Others form organized societies. Living in a group is very helpful. Together, animals are stronger. They are better able to hunt, feed, raise their young, defend themselves, and survive in the wild.

Hunting as a pack

If an animal intends to hunt something bigger than itself, it should join forces with others. Like lions and wolves, African hunting dogs form hunting packs. These wild dogs roam African savannahs in groups of about twenty. They pursue and harass gazelles, wildebeests, and impalas until their exhausted prey surrenders.

African hunting dog

Are you curious?

In the animal world, the strongest individuals are often the first to feed. Among African hunting dogs, however, the youngest and the oldest are allowed to eat first.

Community incubation

Among several bird species, each member of the group works together to make sure the young get the care they need. The white-winged chough builds a community nest. Two females then lay all the eggs — about ten. The entire group then watches over and sits on the eggs until they hatch 13 to 25 days later.

white-winged chough

chaffinch

All for one and one for all!

Many species of small birds work together to defend themselves against birds of prey. Some of these small birds, such as the chaffinch, perform amazing sky dances. They whirl together in the air and make headlong charges until their enemies get dizzy and are forced to flee.

A community trap

Tiny Mexican spiders are barely 0.2 inch (5 millimeters) long. They gather in colonies of about one hundred. These little insects work together to build a huge web around the branches of a tree. This trap even includes little rooms. It can ensnare prey large enough to feed the entire colony at one sitting!

communal spider

The comforts of family life

In nature, males and females must come together to reproduce. Often, one parent is left alone to bring up the little ones. However, some animals stay together and care for their young, who have to be watched over and fed until they are old enough to fend for themselves. Some animals form new couples at the beginning of every mating season. Other couples stay together forever.

Lifelong couples

Male and female swans remain with each other for their entire lives. Both whooper swan parents help raise their young. The birds always travel on water as a family. Led by their mother, the young swans sometimes climb onto her back to rest on her soft feathers. The father watches over his offspring from the rear, warding off danger.

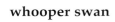

whooper swan

Are you curious?

Geese, ducks, and swans belong to the same family. Swans are different from geese because they have longer necks, which allow them to feed in deeper water.

A joyous family

The Australian laughing jackass, or kookaburra, makes a giggling sound in the wee hours of the morning. Kookaburras form large communities. Young kookaburras are able to reproduce, but they stay with their family for the first few years of their adult life. They don't leave until it's time to start families of their own.

Australian laughing jackass

Brothers and sisters together

Dwarf mongooses live together with their brothers and sisters. For as long as their parents can still bear young, the offspring stay in the group. They help one another care for the newborns and keep the family safe. Even females who have never given birth can produce milk to feed their little brothers and sisters!

dwarf mongoose

bat-eared fox

5

Long live the family!

The bat-eared fox is an African mammal with big ears and silvery fur. Its family is made up of a lifelong couple and offspring of all ages. Young bat-eared foxes become adult-sized in only six months. But they stay with their parents to help them raise the next generation of cubs.

Harems

Few animals live in small family units. Most form larger groups. These groups unite to protect themselves against predators, to find food, to care for their young, and to control their territory. Some societies are formed mainly to reproduce. One male surrounds himself with several females. These groups are known as harems. During the mating season, the male harem leaders gather a group of females that can produce offspring. They mate with these females and defend their territory against other males.

Societies of the African plains

Burchell's zebra, or the common zebra, forms harems made up of one male, six mares, and their foals. When they are old enough to reproduce, the young females leave the group to find a male that wants to enlarge his harem. The young males live in separate groups until they can form their own "societies."

Burchell's zebra

And the winner is . . .

When spring comes to the Arctic, elephant seals climb ashore from the icy northern waters. They form harems that last through the mating season. The males fight violently, and the one that wins the most fights claims the most females! Average males attract between 10 and 20 females. The strongest have harems of about 40 or as many as 100!

northern elephant seal

Reproduction denied!

chuckwalla

Most reptiles lead solitary lives. However, the chuckwalla, a large South American lizard, forms a sort of harem. One male has absolute power over a very large territory. Other males are allowed to live there, but only that strongest male can mate with females. During the mating season, he sends away the other males!

Nordic harems

Caribou enjoy group living. They like the safety a group life provides. These large animals migrate in herds of up to 50,000! When it is time to reproduce, each male caribou forms a temporary harem of 10 to 15 females. After they give birth, the new mothers leave the harem with their fawns and rejoin a group of females.

woodland caribou

Are you
curious?

Each species of zebra has its own pattern of stripes. Each individual zebra has special markings. These differences help zebras recognize each other.

Female societies

In many animal species, the females are the leaders. That is why some animal societies revolve around mothers that are the heads of families. These female societies are often made up of sisters, their daughters, and their daughters' offspring. These groups of sisters, aunts, and cousins form matriarchal societies where cooperation and mutual support are the order of the day!

Female mountain climbers

Female Rocky Mountain goats travel in groups that include their offspring, called kids. The males form separate groups or live alone. Males join females only during the mating season. The females, or nanny goats, totally control their territory. Adult nanny goats show their power by delivering firm blows with their horns.

Are you curious?

During the mating season, male mountain goats dig holes. They urinate in the hole and then roll around in it. The dirtiest, smelliest males seem to have the best chance of finding a mate!

ring-tailed coati

Large female clans

The ring-tailed coati is a rodent, a member of the rat family. Coatis form groups of 20 to 50 or more females and their offspring. The males live alone. They join the group only during the mating period and do not let any other males join the group. Even males barely two years old are driven away!

white whale

Marine societies

White (beluga) whales form groups of about fifteen females and their offspring. Baby beluga whales stay with their mothers for several years. Females do not reach adulthood until they are about five years old. Males do not reach maturity until about the age of eight. They eventually leave their mothers to form separate groups of males.

African elephant

A leader of the herd

The largest land animal, the elephant, lives in a society called a herd. The herd is made up of many families — adults and their young. A herd of African elephants may number up to 1,000! The animals help protect one another and care for the baby elephants. Most herds are led by one of the oldest females.

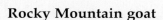

Rocky Mountain goat

Social orders

Some animals create orderly societies. They are often based on the power of certain animals over others. In these societies, adults usually rule the young and males have power over females. Rankings are usually decided by battles. Those who are victorious get the best food and the largest number of female partners. But they also have to protect their society and find food for all its members.

grey wolf

Young leaders

Social rank in a wolf pack is decided early. Cubs fight each other to determine their place in society. A head couple rules its offspring, and several relatives make up the rest of the pack. The head male leads the pack, organizes hunts, and deals with enemies. Following in the social order are the other males, his female partner, the other females, and then the young wolves.

Pecking order

Hens live by special rules. Two by two, they peck at each other. The winner of each battle takes on another winner until there is a champion. Each hen takes her place in the pecking order based on the fights she has won. The hens highest up may express their standing in the group by continuing to peck away at their inferiors.

domestic hen

Social marsupials

The whiptail wallabies of Australia live in communities of thirty to fifty. Within the community, smaller temporary groups come together to feed or to rest. The adult males are usually peaceful, but they often fight over the females. The results of the fights determine which male may mate with the females.

whiptail wallaby

jackdaw

A society of birds

Jackdaws are closely related to ravens and crows. They live in tree cavities, walls, ruins, and steeples. The rules of jackdaw society are established by fights at the beginning of each season. The rules are strict: no male can mate with a female whose social rank is higher than his own, and each female is given the social rank of her mate.

Are you curious?

The expression "hungry as a wolf" reflects the eating habits of these canines. In fact, they can wolf down up to 22 pounds (10 kilograms) of meat each day! But they can also live on absolutely nothing for several days in a row.

Primate societies

Apes are skillful society builders. The types of societies they build depend on their lifestyles. Primates that live in treetops have little to fear from predators, so they travel alone or in small groups. Baboons and chimpanzees live and feed on the ground, where they are more likely to be attacked. To protect themselves, they gather together and help each other. Their tribes are often made up of about 100 individuals.

Highly organized societies

Ring-tailed lemurs live in the mountains of Madagascar. In groups of about twenty, these little primates raise their young together. Their groups are highly organized and are governed by strict rules. Although outnumbered by the males, the adult females in the group — especially the oldest ones — are its most highly respected members.

ring-tailed lemur

Are you curious?

One of the favorite pastimes of the ring-tailed lemur is group sunbathing. With their outstretched arms and legs reaching for the sun, they soak up the hot rays. What better way to relax after a meal?

Peaceful families

Gibbons form small, tightly knit family units. The mother and father defend their territory by biting any other gibbon that gets too close. The parents have only one baby about every two or three years. Since the babies do not reach adulthood until the age of seven, the families — with two to four offspring — remain together for quite some time.

white-handed gibbon

Solitary apes

Orangutans live in the trees. The females and their last-born babies spend their days searching for the fruit, leaves, pieces of bark, and birds' eggs on which they feed. The males live alone. During the mating season, they join small groups of females.

orangutan

13

orabussu titi

Families in the trees

Barely 12 to 16 inches (30 to 40 centimeters) long, the orabussu titi lives in the rain forests of South America. The small families of these primates are made up of a male, a female, and one or two offspring. They always travel as a group. The parents are lifelong partners. They occupy and defend an area about 55 yards (50 meters) in diameter.

More primate societies

Different species of primates live in different kinds of societies. Some live alone; some live as couples; some have harems; and some live in troops that share the same territory. The groups formed by primates are usually very stable, and the bonds among them are very strong. They communicate in many ways — with cries, grooming sessions, caresses, and facial expressions.

A wise animal with a silver back

The gorilla is the largest living primate and one of the most intelligent land animals. These giants of the African forests live in small harems. A single adult male watches over a group of females and their young. This caring and protective father is known as a silverback because of the white hair on its back. The silverback is responsible for the entire group.

14

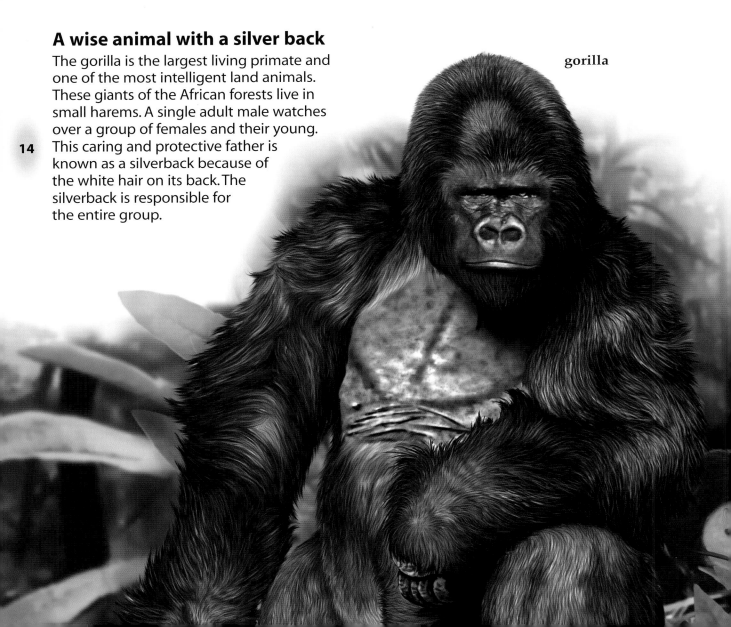

gorilla

Community life

Common chimpanzees live in groups made up of as many as 100 males and females. Smaller groups may be formed within these communities. Some of these groups include mothers and their babies. Others are made up of males who defend the community's territory. Male chimpanzees dominate females, and older females dominate younger ones.

common chimpanzee

Harems headed by women

Red guenons roam the African plains in groups of five to thirty. Although they live in harems, the females are the leaders of the red guenon community. A female leads the group, and the females have the last say when conflicts arise. The males are responsible only for protecting the group.

red guenon

hamadryas baboon

Ruling males

Among hamadryas baboons, the smallest family unit is a harem made up of a possessive male and the five to ten females that are at his beck and call. Several harems often unite to form clans of ten to twenty baboons. Together, these hamadryas clans form troops of about seventy animals.

Are you curious?

Only a few thousand gorillas are left in the wild, and this number continues to go down. Human beings cut down the African forests where gorillas live, hunt them for food, and capture them for zoos.

Special roles to play

Some animals have highly organized societies. In these groups, no individual could survive on its own! The lives of social insects are strictly organized. Each individual in the group, or colony, has a role. Those with the same role – kings, queens, soldiers, workers, or drones – look identical and are divided into groups known as "castes." Whatever their role, all members of a colony perform tasks to ensure the survival of the group as a whole.

A queen's life

Of the 50,000 to 80,000 bees living under the same roof, just one female – the queen – is responsible for reproduction. About 100 males, called drones, mate with the queen. Tens of thousands of worker bees take care of everything else. They build, clean, and repair the cells; they feed the larvae; and they gather the nectar and store it in the cells of the hive.

honeybee

A caste society of mammals

Naked mole rats are rodents with pink, bare, wrinkled skin that live in colonies of about forty. They are divided into castes, just like colonies of social insects. The queen rules small male and female workers and large soldiers. The workers build an impressive network of tunnels, and they find food. The soldiers, who defend the tunnels, spend most of their time snoozing!

naked mole rat

termite

A termite monarchy

Two million species of termites inhabit the Earth. They form colonies of up to a million each! Each colony revolves around a couple — the king and the queen. They are the only ones allowed to reproduce. The big-jawed soldiers defend the colony. The workers provide food for all, and they build and maintain the termitarium.

African stink ant

The birth of an ant hill

Ant societies are divided into three castes: one or more queens, the males, and the female workers. Every so often, one queen and one male leave their colony to start a new one. They produce males to continue reproduction and workers to care for the queen and to clean and defend the ant hill. The males die shortly after they mate with a queen.

Are you curious?

The queen bee is a baby factory. She lays 1,500 to 2,000 eggs per day. Because she can live four to five years, she may lay more than 2 million eggs over the course of her lifetime!

Joining the crowd

Just because animals live in groups does not mean they form societies! Some animals cannot even recognize each other, but they prefer group living. Attracted by a source of light or food or by the need for protection, they cluster together in huge groups. This explains why herds of mammals, flocks of birds, swarms of insects, and schools of fish migrate, hibernate, sleep, rest, reproduce, or just move around in vast numbers.

marine iguana

Sunbathing iguanas

In the Galapagos Islands, thousands of marine iguanas stretch out on the rocks to warm themselves in the sun after a long, chilly night. The 3-foot- (1-m-) long members of these herds of reptiles go down to the sea only to feed. Then they return to their restful perches on the rocks.

Black clouds of locusts

Desert locusts can gather together to form swarms of several million insects. Traveling thousands of miles (kilometers) in search of food, these black clouds of locusts eat hundreds of pounds (kg) of vegetation every day. They can destroy millions of dollars worth of crops.

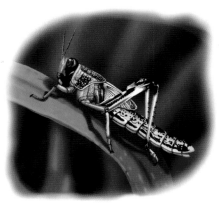

desert locust

Everyone all in a row

Processionary caterpillars move around one after the other. Pine processionary caterpillars, which are usually found in pine forests, spend most of their time together. They travel in groups of up to 300, forming a train 33 feet (10 m) long. Each caterpillar follows the tuft of hair at the end of the preceding one.

pine processionary caterpillar

Norway lemming

Mass migrations

Every few years, Norway lemmings bear great numbers of young. The females give birth to as many as eight young at a time, and they do so several times a year! When their territory becomes too crowded, these little rodents flock toward fertile valleys in search of new spaces to live and feed.

Are you curious?

The marine iguana is the only marine lizard. When the tide reaches its low point, every three to five days, this lizard dives for the seaweed on which it feeds. It can dive as deep as 50 feet (15 m)!

Partners

Certain animals that belong to different species form close bonds based on partnership and cooperation. These "associates" may unite permanently or only occasionally. There are three main types of partnerships. In one type, one animal benefits without bothering the other partner. In another type, one animal uses another in a harmful way. In the third type, both animals benefit.

Comforting arms

Sea anemones have a powerful weapon. Concealed in their tentacles are cells that can paralyze and kill any prey that comes near. The little clown fish found on coral reefs spend most of their time in and around the stinging cells. They even raise their offspring there! The skin of this fish has a protective mucus that allows it to hide in an anemone at any time.

clown fish

striped remora

Fast transportation

The remora is a fish that travels without using up much energy. Instead of a main dorsal fin, it has a sucker that allows it to attach itself to the belly of a shark. Once stuck to its host in this way, the remora can travel many miles (km). It benefits from the protection of the shark and feeds on the shark's parasites.

Turtle transport

Barnacles feed by filtering out the microscopic animals suspended in water. These little crustaceans often attach themselves to rocks or boats. They also like to travel on the backs of turtles, snakes, fish, and marine mammals. By covering great distances, they can find a better food supply.

barnacle

A harmful partner

Ticks make life hard for many poor mammals. After traveling through the fur of their victim, these mites pierce the skin. Then they attach themselves and suck the blood of their host for several hours or even several days. Ticks sometimes carry serious diseases.

tick

Are you curious?

The anemone benefits in several ways from its relationship with the clown fish. The clown fish removes the refuse that collects on the tentacles of the anemone. It also attacks enemies and attracts other small fish on which the anemone feeds.

Side by side

Inseparable animals often have their very own living space, and each can come and go as it pleases. But certain associations are so helpful that where one partner is found, the other is rarely far behind.

ratel

A taste for honey and beeswax

One has a weakness for honey, and the other likes beeswax. It's a match made in heaven! When a greater honeyguide finds a beehive, it cries out until it attracts a ratel. When it arrives, the ratel feasts on the delicious honey. Its partner patiently waits to enjoy the remaining honey and wax.

22

Are you curious?

The skin of the ratel is so thick that bees can't sting it and poisonous snakes such as cobras can't bite it.

A good neighborhood

South American yellow-rumped caciques choose the location of their colony wisely. Pressed up against each other, their little nests are built in large trees overhanging streams. They are often found near wasp nests. This location provides the birds with a measure of protection and does not harm the wasps.

yellow-rumped cacique

common wood ant

large blue caterpillar

One good turn . . .

Although they feed on other things, wood ants love the liquid secreted by the large blue caterpillar. The ants imprison the large blue in the ant hill and feed on the liquid. In return, the caterpillar feeds on the ant brood. Once its larva days are over, a full-fledged butterfly leaves the ant hill.

tsetse fly

A murderous fly

Some harmless-looking flies carry serious diseases. By sucking the blood of certain wild mammals, the tsetse fly ingests dangerous microbes. It can transmit sleeping sickness to domestic animals or human beings. In humans, sleeping sickness is a very serious illness that can cause death.

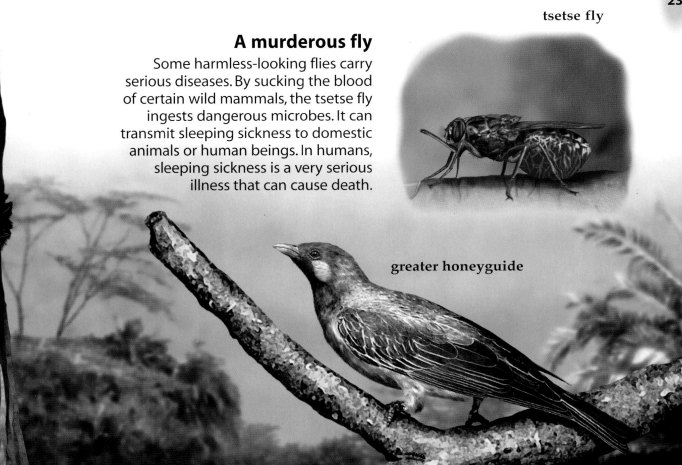

greater honeyguide

A solitary life

Whether they live in families, societies, or herds, most animals are drawn to one another. Some, however, prefer to live alone. Mating season is the only time these animals have contact with other members of their species. As soon as they have mated, the parents separate. One raises the young alone. The animals isolate themselves in an area they aggressively defend against intruders.

A solitary stroller

The black rhinoceros is a large, solitary animal that has little use for its fellow creatures. During the mating season, it has to be very patient to interest a partner. The parents remain together for only a few days, then separate. Each returns to its own territory.

black rhinoceros

A merciless loner

The small triangular head of the European mantis is nearly invisible in the grass. This hunter does not like to be disturbed as it stalks its prey. It even eats other mantises that dare to enter its living space — even if the intruder is its own mate!

European mantis

An arctic loner

The male polar bear has little to do with other polar bears. He gains the right to mate by engaging in numerous fights with his rivals. The victor does not stay with his mate, however. He abandons her to care for the cubs and keep the family together on her own.

polar bear

okapi

Mother and "child"

A member of the giraffe family, the strange-looking okapi is a loner that lives in the forests of central Africa. The female okapi and its most recent baby get together with a male only when it is time for the female to mate. These solitary animals use their sense of smell to locate each other.

25

Are you
curious?

The last Indicotherium, the ancestor of today's rhinoceros, lived 10 million years ago. This giant creature weighed 33 tons (30 metric tons) and measured 20 feet (6 m) at the withers. It had a long neck that allowed it to reach leaves growing 26 feet (8 m) above the ground.

A map of where they live

More fun facts

BRIEF GLOSSARY OF ANIMAL GROUPS		
Name of Group	**Definition**	**Examples**
Herd	A group of wild or domestic animals of the same species	Wild horses, buffalo, deer, boars, cattle, sheep
Flock	A group of wild or domestic animals of the same species	Sheep, goats, seagulls
Harem	A group of several females gathered around a male for the purpose of reproduction. The harem is fiercely defended by the male	Musk-ox, sea lion, hippopotamus
Pack	A group of animals that live together, especially predators and hunting animals	Coyotes, wolves, African hunting dogs, hounds
School	A large number of fish of the same species	Herring, cod, tuna
Cloud	A large number of insects or other small animals in flight	Birds, mosquitos
Swarm	A large mass of small animals, especially insects	Bees, migrating locusts
Rookery	A colony of birds or mammals such as seals or penguins	Sea lion, emperor penguin
Colony	A group of animals that live together, such as a group of birds that gather together to reproduce, or a group of stationary animals such as coral	Gannet, bees, coral
Troop, band	A herd or flock	Monkeys
Community	An animal population that makes up a group in the same location	Fish, amphibians, birds, or mammals that inhabit a particular lake
Society	A group of individuals of the same species that move together, recognize each other, communicate with each other, and follow common rules	Wolves, elephants, most primates
Mass	A large gathering of animals with no social organization	Insects around a light source, migrating butterflies

HOW MANY ARE THERE?		
	Animal	**Number**
Loners	Dragonfly	1
	Snake	1
	Turtle	1
	Marten	1
	Leopard	1
Families	Gibbons	3 to 4
	Swans	4 to 9
	Beavers	5 to 12
Groups	Hippopotamus	10 to 15
	Blue hare	15 to 30
	Japanese macaques	30 to 40
	Capybaras	65
	Hyenas	100
Multitudes	Ants (Formicidae family)	10,000 to 100,000
	Honeybees	50,000 to 80,000
	Red-billed quelea (birds)	10 to 30 million
	Monarch butterflies	16 million
	Free-tailed bats	20 million
	Black-tailed prairie dogs	400 million
	Desert locusts	250 billion
	Rocky Mountain locusts	12,500 billion

African hunting dog

size	4.5 feet (1.36 m) including the tail; 24 inches (60 cm) at the withers
weight	35 to 60 pounds (16 to 27 kg)
distribution	Africa
habitat	savannahs, plains, semi-arid regions
diet	gazelles, impalas, zebras, gnus, antelopes
reproduction	2 to 12 offspring per litter; 9- to 11-week gestation period
life span	10 to 15 years

class	Mammalia
order	Carnivora
family	Canidae

whooper swan

size	5 feet (1.5 m) long; wingspan: 6.6 feet (2 m)
weight	about 24 pounds (11 kg)
distribution	the Arctic, Iceland, Scandinavia, Siberia, Europe
habitat	fresh water, marshes, lakes, bays
diet	aquatic plants, worms, molluscs, insects
reproduction	from 5 to 7 eggs, in May; 35- to 42-day incubation period
life span	up to 30 years

class	Birds
order	Anseriformes
family	Anatidae

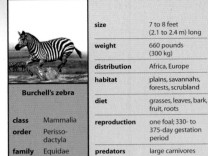

Burchell's zebra

size	7 to 8 feet (2.1 to 2.4 m) long
weight	660 pounds (300 kg)
distribution	Africa, Europe
habitat	plains, savannahs, forests, scrubland
diet	grasses, leaves, bark, fruit, roots
reproduction	one foal; 330- to 375-day gestation period
predators	large carnivores

class	Mammalia
order	Perissodactyla
family	Equidae

Rocky Mountain goat

size	about 3 feet (1 m) at the withers
distribution	western Canada, the northwestern United States
habitat	steep slopes, alpine meadows
diet	grasses, shrubs, flowers, coniferous buds, moss, lichen
reproduction	one kid; 178-day gestation period
predators	cougar, eagle, brown bear, wolverine, wolf, coyote

class	Mammalia
order	Artiodactyla
family	Bovidae

grey wolf

size	3 to 5 feet (1 to 1.5) m long
weight	165 pounds (75 kg)
distribution	North America, Asia, Middle East, Europe
habitat	open areas, forests, ice fields
diet	deer, carrion, small mammals, insects, fruits, vegetables
reproduction	2 to 8 offspring per litter; 61- to 63-day gestation period
life span	about 12 years

class	Mammalia
order	Carnivora
family	Canidae

ring-tailed lemur

size	15 to 18 inches (39 to 46 cm) including the head; tail: 22 to 25 inches (56 to 63 cm)
weight	5 to 7 pounds (2.3 to 3 kg)
distribution	Madagascar
habitat	forests and open, dry areas
diet	leaves, fruit, seeds
reproduction	one offspring per litter; 136-day gestation period
predators	birds of prey
life span	15 to 18 years in captivity

class	Mammalia
order	Primata
family	Lemuridae

gorilla

size	from 5 to 6 feet (1.5 to 1.8 m) tall
weight	200 to 400 pounds (90 to 180 kg)
distribution	central and western Africa
habitat	tropical forests
diet	plants, bushes, vines
reproduction	one baby gorilla per litter; 250- to 270-day gestation period
predators	humans
life span	35 years in the wild

class	Mammalia
order	Primata
family	Pongidae

honeybee

size	less than 1 inch (25 mm)
distribution	throughout the world
habitat	open areas with an abundance of flowers
diet	nectar and pollen
reproduction	1,500 to 2,000 eggs per day
predators	birds, wasps, dragonflies
life span	queen: 5 years; drones and workers: 1 month

class	Insecta
order	Hymenoptera
family	Apoidea

marine iguana

size	up to 6 feet (1.8 m) long
weight	7.5 pounds (3.4 kg)
distribution	Galapagos Islands
habitat	coasts
diet	seaweed and kelp
reproduction	2 eggs; 4-month incubation period
predators	wild dogs

class	Reptilia
order	Squamata
family	Iguanidae

clown fish

size	4.3 inches (11 cm)
distribution	western and central Pacific Ocean
habitat	coral reefs
diet	microscopic plants and animals
reproduction	20,000 to 25,000 eggs at a time
predators	fish
life span	over 5 years

class	Fish
order	Percomorphi
family	Pomacentridae

ratel

size	31.5 to 35.5 inches (80 to 90 cm), including the tail
weight	about 24 pounds (11 kg)
distribution	Africa, Saudi Arabia, Asia, India
habitat	rock hills, forests, savannahs, plateaus
diet	fruit, rodents, birds, insects
reproduction	2 young per litter; 6-month gestation period
predators	large carnivores
life span	up to 20 years

class	Mammalia
order	Carnivora
family	Mustelidae

black rhinoceros

size	10 to 12 feet (3 to 3.7 m); 5 feet (1.5 m) at the withers
weight	1 to 2 tons (1 to 1.8 metric tons)
distribution	Africa
habitat	wooded savannahs
diet	plants, especially acacia branches
reproduction	one calf per litter; 460-day gestation period
predators	lions (hunt the calves)
life span	40 to 50 years

class	Mammalia
order	Perissodactyla
family	Rhinocerotidae

Glossary

abandon: To leave without intending to return

brood: Batch of eggs or larvae produced by animals such as bees

canine: A member of the dog family

crustacean: Any of a class of animals that have shells and many pairs of legs

dominate: To have authority over others

dorsal: Located on or near the back

family unit: Group of individuals forming a family

fawn: Young of the deer or of a related species

fertile: Capable of producing offspring

generation: Set of beings born at about the same time that are about the same age

hibernate: To spend the winter in a resting state

incubation: The process of sitting on eggs to warm and hatch them

ingest: To take in for digestion

isolate: To set apart from others

larva: An often wormlike form that is one stage of an animal's life

mammal: A member of any animal species in which the female has mammary glands for feeding her young

marsupial: An animal whose young spend several months after birth in their mother's pouch, where they nurse from her mammary glands

matriarchal society: A society in which females have the decision-making power

maturity: Age or period of life when an animal completes its development and becomes an adult

microbe: A microscopic organism; a germ

migration: A mass movement of a single species at a specific time of the year and in a specific direction

mucus: Thick, transparent liquid

nectar: A sweet liquid produced by certain plants

offspring: Child, young animal

paralyze: To make powerless, unable to move

parasite: A living organism that lives on or in, and gets nourishment from, a host organism; often harmful to the host

pattern: A design

possessive: With a desire to own or dominate

predator: An animal that destroys or eats another

prey: An animal that is the victim of a predator

primate: An order of mammals that inclues human beings, apes, and monkeys

reproduce: To produce new individuals of the same kind

reptile: A crawling animal with scale-covered skin, such as the snake, the iguana, and the tortoise

rodent: A mammal with sharp incisors that eats by gnawing, such as the mouse

savannah: A treeless plain

secrete: To form and give off from the body

social rank: Position occupied by an individual in a group in relation to the other members

society: A cooperating social group whose members interact with one another

solitary: Alone, without companions

species: A class of related organisms

stinging cell: A cell that discharges an arrowlike barb and a toxin that stuns prey

store: To set aside for future use

tentacle: An elongated flexible arm, often lined with suckers, used by certain mollusks to touch and grasp.

termitarium: A termites' nest

territory: An area an animal reserves for itself, forbidding access to others

withers: The highest part of the back of a quadruped

Index

Editorial Director Caroline Fortin **Research and Editing** Martine Podesto **Documentation** Anne-Marie Brault, Anne-Marie Labrecque **Page Setup** Lucie Mc Brearty **Illustrations** François Escalmel, Jocelyn Gardner **Translator** Gordon Martin **Copy Editing** Veronica Schami **Gareth Stevens editing** Joan Downing **Cover Design** Joel Bucaro, Scott Krall